FIRE ENGINES

Eddie Baker

SHIRE PUBLICATIONS

Bloomsbury Publishing Plc

PO Box 883, Oxford, OX1 9PL, UK

1385 Broadway, 5th Floor, New York, NY 10018, USA

E-mail: shire@bloomsbury.com

www.shirebooks.co.uk

SHIRE is a trademark of Osprey Publishing Ltd

First published in Great Britain in 2018

ISBN: PB 978 1 78442 300 1
         eBook 978 1 78442 299 8
         ePDF 978 1 78442 298 1
         XML 978 1 78442 301 8

18 19 20 21 22   10 9 8 7 6 5 4 3 2 1

Typeset by PDQ Digital Media Solutions, Bungay, UK

Printed in India by Replika Press Private Ltd

Shire Publications supports the Woodland Trust, the UK's leading woodland conservation charity. Between 2014 and 2018 our donations are being spent on their Centenary Woods project in the UK.

COVER IMAGE
Front cover: A 1952 Dennis F15 water tender from Glamorgan Fire Service. Back cover: iStock.

TITLE PAGE IMAGE
This Tilley manual fire engine represents the rapid development of horse drawn manual fire engines in the early nineteenth century.

CONTENTS PAGE IMAGE
Built in 1966 by Hampshire Fire Brigade's workshops was this Foam Tanker based on a Bedford TK chassis. It carried ready mixed foam but had no built in pump.

ACKNOWLEDGEMENTS
Images are acknowledged as follows:

Pete Ashpool Collection, pages 9 and 15 (top); Jim Bavin, page 31; Mike Bunn Collection, page 24; Colin Carter, pages 48, 52, 53; Gary Chapman, pages 42, 45 (bottom), 46 (bottom), 56 (top); Colin Dunford, pages 47 and 49 (top); Neil Fraser, page 16; Getty Images, pages 1, 4, 5, 10, 11, 23 (top); Keith Halstead, page 32; John Shakespeare, pages 3, 13, 17, 22, 26, 28, 29, 30, 33, 34 (top), 34 (bottom), 35, 36, 38, 39 (top), 40 (bottom), 41 (bottom), 58 (top), 58 (bottom), 59 (top); Mike Sudds, pages 50, 51 (top), 51 (bottom).

All other images are in the author's collection.

CONTENTS

EARLY DAYS 4

1900–1920 9

1921–1939 14

THE SECOND WORLD WAR 22

1945–1959 26

THE 1960s AND 1970s 35

THE 1980s AND 1990s 42

THE TWENTY-FIRST CENTURY 48

AIRPORTS AND INDUSTRIAL BRIGADES 57

GLOSSARY 60

PLACES TO VISIT 62

FURTHER READING 63

INDEX 64

EARLY DAYS

Since man discovered fire there has always been a problem with fire spreading to devour both life and property. The Romans had an organised firefighting force by 300 BC and by the first century had established the Corps of Vigiles. Their equipment would have been very basic, consisting of long hooks and axes, which were used to pull down buildings to stop the spread of fire, and bucket chains to throw water on to the fire. When the Roman Empire collapsed so did their expertise in firefighting.

As urban living became more popular the risk of destruction became very high and there were many instances

Richard Newsham manual pump from 1734. This example came from Dartmouth and has two single-action pumps and an air vessel placed in the tank.

of whole villages and small towns being destroyed. It was not until the sixteenth century, however, that the use of 'squirts' came into use. These were made of brass and took one or two men to hold the unit and a third to pull and push the piston handle. It was the Great Fire of London in 1666 that raised the nation's interest in any organised form of fire protection and the insurance companies set up their own fire brigades.

Horse-drawn manual pump built by Merryweather and Sons in 1866. With twenty-two men to work it the engine could deliver 100 gallons of water per minute to a height of 120 feet. This example was purchased by the Duke of Portland for his Welbeck Abbey Estate.

By the seventeenth century, manually operated pumps had come into use. The most common design consisted of a pump mechanism housed in a cistern which was filled by a bucket chain. The pump was operated by four or five men who worked a handle up and down, forcing the water out through a nozzle fixed to the top of the unit. Later designs could draw water through a suction hose and pump it out through a leather hose, which could be taken closer to the fire.

In 1721 Richard Newsham produced a pump that out-performed all previous designs and now produced a constant jet of water. Newsham's successful design was soon being adapted and copied by fire engineers worldwide.

The design of these engines became larger and larger and by the beginning of the nineteenth century the use of horse-drawn engines became a natural progression.

The concept of the different insurance brigades dealing with fires only in their own insured buildings led to intense rivalry which was not helpful to efficient firefighting. In London this eventually led to the amalgamation in 1833 of the insurance brigades into the new London Fire Engine Establishment (LFEE). With thirteen fire stations this was still a private fire brigade funded by the insurance companies and as such was mainly responsible for saving goods from fire.

The first steam fire engine was built in 1829 by John Braithwaite of London to a Swedish design and was capable of pumping water to a height of 90 feet. Liverpool became the first brigade in Britain to operate a steamer in 1831 and it remained in use for many years. James Braidwood, chief of the LFEE, was strongly against steam-powered engines but relented for the river service and had a manual operated fire float converted to steam power.

The advent of steam power did not herald the end of manual power as by the middle of the nineteenth century Merryweather and Sons were still building horse-drawn manual fire engines requiring from six men to forty-six men to operate them. The largest was said to be able to pump water to a height of 150 feet. With the large number of men required to operate these machines it was obvious that help from onlookers was needed and tokens were handed out which entitled the helpers to free beer.

In the 1850s the name of Shand Mason & Co. came onto the scene and soon established itself as the leader in the field of fire engine design. By 1860 the LFEE had seen the light and was operating a hired Shand Mason engine; three more were placed in service in 1866, later followed by a further ten.

Merryweather & Sons built their first steam fire engine in 1861 and were soon building a number of different designs including the Metropolitan Fire Brigade No. 1 Pattern, which came in two sizes: 350 and 450 gallons per minute (gpm).

A number of large fires including the Palace of Westminster in 1834 and Tooley Street in 1861 had spurred the insurance companies to lobby the government to provide London with a brigade at public expense and management. The Metropolitan Fire Brigade Act of 1865 was passed creating the Metropolitan Fire Brigade (MFB), which in 1904 changed its name to the London Fire Brigade (LFB).

By 1866 there were sixty Shand Mason steamers in service in the UK, seventeen built by Merryweather and sixteen by various other manufacturers. It is apparent that Shand Mason and Merryweather had secured the market with few other manufacturers coming forward.

As water mains improved and fire hydrants became more common in the mid-nineteenth century, some town brigades, having a good water pressure, could connect their hose directly to a standpipe without having to use an intermediate pump to maintain pressure. This avoided the expense of providing a manual or steam pump and instead a hose cart or hosereel cart was used to stow the hose and equipment.

Hose carts were suitable for towns with closely spaced hydrants, which only required small runs of hose. The basic hose cart was a box mounted on a two-wheeled hand-pulled carriage carrying a supply of hose, a standpipe, branch-pipe, nozzles and hand tools.

The hosereel cart, where coupled hose was wound on a reel, was more suitable when a longer run of hose was required. The reel was mounted between the two wheels of the carriage and could carry up to 2,000 feet of hose. An equipment box was invariably fitted across the top of the carriage. Hand hosereels were most common, although larger models were available to be horse-drawn with a seat for two men.

In the early nineteenth century London had a number of street escape stations situated at strategic street corners; these were originally independent of the fire brigade until 1866 when the MFB took full control. Each station consisted of a

This simple hose cart was in use at Sandringham House, Norfolk. Note the fittings for carrying a ladder.

cabin to house the watchman, more commonly known as the conductor, and a wheeled wooden escape ladder. When there was an outbreak of fire the conductor and any available help would quickly push the ladder to the fire to enable rescues to be performed. Horse-drawn escape carriages appeared from 1890 and became known as escape vans.

Hook ladders had been introduced from France and became common in the UK during the 1880s and the first extending ladder was supplied to London in 1890.

Turntable ladders consist of an extending ladder mounted on a turntable, which enables it to be rotated through 360 degrees. The ladder consists of a main section that can be raised to varying degrees of pitch. Within the main section are a number of sliding sections, which can be extended to lengthen the ladder. Shand Mason was advertising a horse-drawn four-wheeled turntable ladder in 1896 and Edinburgh built a 65-foot horse-drawn model that had three sections and was operated manually.

There were frequent delays in firefighting due to the time spent either searching for water or connecting pumps to street hydrants. Towards the end of the nineteenth century the chemical engine was introduced to provide an instant response in firefighting. These engines consisted of a horse-drawn vehicle similar to a manual pump. Cylinders containing some 50 gallons of water into which bicarbonate of soda had been dissolved were mounted on the carriage. Sulphuric acid was released into the container and the contents mixed by agitation. The chemical reaction produced carbon dioxide, which pressurised the cylinder, ejecting the water via a hose onto the fire.

In 1898 Shand Mason introduced a first-response engine built on a four-wheeled horse-drawn carriage which used compressed air to pressurise the water cylinders, with a shut-off valve allowing the water to be shut down. This was an improvement on the soda-acid type, which had to be fully discharged until the cylinders were empty.

1900–1920

THE LATTER YEARS of the nineteenth century had seen various attempts at converting horse-drawn steamers to self-propulsion and in 1902 Liverpool Fire Brigade purchased its first Merryweather Fire King. The steam-powered engine drove the rear wheels and the vehicle weighed nearly 6 tons and ran on solid rubber tyres. It was capable of pumping a jet of water to a height of 150 feet at 350gpm and had a speed of 20mph, making it one of the most powerful fire engines of its time. By 1908 London had six self-propelled Merryweather Fire Kings in service with the last recorded fire attended by a Fire King in London being in July 1912, after which all were returned for part exchange.

London's last horse-drawn steamer, a Merryweather, was delivered in 1906 although at the end of 1912 there were still

A much larger hose cart than that at Sandringham, needing six men to pull it.

The London Fire Brigade's first motor steamer, c.1907. Both the water pump and the vehicle itself are powered by steam, as opposed to the earlier steam pumps pulled by horses.

nearly 450 Shand Mason steamers in use in Britain and as late as 1923 some brigades were still placing orders. It is worth noting that some steamers remained in operation well into the Second World War.

The first petrol-engine appliance recorded was in 1901; this was delivered to Eccles in Lancashire. Others soon followed, mainly consisting of combination appliances carrying escapes and either chemical or air-powered pumps built by such companies as Merryweather and John Morris & Son.

The need for re-breathable air and illumination at incidents soon became apparent and the evolution of the Emergency Tender began. It is believed that the first such vehicle was delivered to Manchester Fire Brigade in 1904: a horse-drawn appliance with a steam-powered air pump and dynamo. The air pump supplied air via hoses to four smoke helmets and the dynamo powered eight electric lights. By 1912 the London Fire Brigade was operating a petrol-engined Dennis Emergency Tender carrying oxygen-breathing apparatus, with searchlights powered by a built-in generator.

Before the widespread use of petrol engines some large brigades experimented with the use of battery-powered electric engines for a short period (around 1905) but these

Horse-drawn escape vehicle, 1910. Firemen demonstrating a horse-drawn escape vehicle at the Southwark Headquarters of the London Fire Brigade.

were very heavy due to the large number of batteries required. This design was pioneered by a French company – Cedes – but the heavy batteries lasted only a short time, just enough for one call-out of no more than 20 miles.

With the introduction of motor-driven pumps manufacturers were able to fit water tanks supplying hosereels and the chemical engine began to disappear. They had given excellent service, albeit over a relatively short period, and by 1910 had all but disappeared.

It was obvious that the introduction of the petrol engine would see a change in fire engine design. First response engines and escape vans had proved successful and pumps would soon follow. The pump itself was either a reciprocating pump, the same as those used by steamers, or the newly designed centrifugal pump. The centrifugal pump consists of a casing containing an 'impeller' with vanes, which rotates at great speed. As water is introduced into the centre of the casing it is propelled outwards by centrifugal force and discharged through the outlet at pressure. The centrifugal pump is still the most widely used pump in modern firefighting.

In 1908 the first Motor Pump fitted with a centrifugal pump was delivered to Bury in Lancashire. It was also fitted with a small auxiliary pump supplying a hosereel taking water from a built-in water tank.

Other manufacturers soon followed, with Dennis Bros of Guildford supplying their first fire engine to Bradford in 1908 designated as the 'N' type. All N types used White & Poppe engines and were fitted with Gwynne pumps. London commissioned its first motor pump from Dennis Bros in 1911.

Meanwhile Leyland Motors delivered an appliance to Dublin in 1909 with a four-cylinder engine capable of a top speed of 60mph and fitted with a single stage pump.

In 1913 Shand Mason demonstrated a steam pump mounted on a Daimler car chassis. This had room for men and equipment including an extending ladder. A number of brigades followed suit and installed a steam pump on a motor chassis to take advantage of the petrol engine without the expense of buying a complete new appliance.

At the time of the outbreak of the First World War Dennis Bros began to dominate the UK market, although Albion, Bedford, Commer, Leyland and Merryweather were all producing fire engines in significant numbers.

The line-up at Croydon Fire Station in 1920 from left to right: a Motor Pump, a Motor Pump Escape, another Motor Pump and two ambulances.

During the First World War the Army requisitioned hundreds of horses, leaving many fire brigades unable to call on their normal suppliers. As a consequence, both manual and steam fire engines were often towed behind motor vehicles.

To combat the risk to the capital from air raids, a special scheme to give the London Fire Brigade support involved

outer London brigades sending appliances into London on request; these were the first organised attempts at mutual support.

White & Poppe, engine manufacturers, had expanded during the First World War but struggled to survive afterwards. In 1919

Typical of the early appliances was this Dennis Pump supplied to St Austell. The crew sat on top of the Braidwood-style body, which offered little protection against falling off the machine, often resulting in fatalities.

the company was taken over by Dennis Bros, production transferred to Guildford, and soon the name was dropped.

The early years of the twentieth century had seen developments in the operation of Turntable Ladders with motor-powered examples. Ladders were now available between 60 feet and 120 feet powered by gas, compressed air or a separate petrol engine mounted on the turntable. By 1919 the LFB were operating four Cedes battery-powered ladders and two Tilling-Stevens ladders. The last horse-drawn appliance, a Turntable Ladder, was withdrawn in 1921.

The design of the early motor fire engines followed that of the horse-drawn era with a hose and equipment box built behind the front seats. The crew sat or stood on this and it was known as the 'Braidwood style' – named after James Braidwood the nineteenth-century chief fire officer. It is perhaps not surprising that many firemen were thrown off, many with fatal results, while trying to dress in their firefighting kit. It should be remembered that horse-drawn appliances had only travelled at about 10mph, while motor engines were travelling much faster, up to 50mph, at which speed the open bodywork was unsuitable. Nevertheless, open bodywork remained a popular choice up to the late 1930s.

1921–1939

ALTHOUGH FIRE BRIGADES were in a state of transition from horse-drawn to motor pumps there was no change in the organisation of brigades, with only London having any obligation to provide a fire brigade. There were no standards of efficiency or co-operation for support and equipment was not built to any standard pattern; for example hose couplings and hydrant outlets.

Crew safety began to improve, however, with the introduction of the New World style bodywork, where the crew sat on inward-facing seats down the sides of vehicles. An alternative design utilised tranverse seating where the crew sat on a forward-facing seat behind the driver and officer.

This immaculate Leyland FK4 Cub Pump Escape registered BYC 145 was delivered to Yeovil Fire Brigade in 1936 and is now in preservation.

Where larger crews had to be carried, the New World body style was introduced, where the crew sat on seats within the body facing each other. This Dennis Ace Pump was supplied to St Neots UDC in 1934.

Both designs provided greater safety and comfort although firefighters were still exposed to the weather.

The year 1931 saw the introduction of limousine bodies affording even better safety and protection for crews. Merryweather built the first for Edinburgh on an Albion chassis, which incorporated a saloon body with seating for twelve men down the sides of the interior. Equipment was stored in lockers under the seats and access was by doors at the rear. Later that year Dennis Bros delivered its first

One design to improve safety was the transverse seating design, where the crew sat on a bench seat behind the officer and driver. This Leyland Pump Escape was delivered to Bristol Fire Brigade in 1937. It did not have the safety feature of cab doors.

This Leyland FK Pump Escape JM 4779, of Windermere Fire Brigade, had transverse seating and cab doors but still no weather protection

limousine-bodied appliance to Darlington; this was better designed, as all equipment was accessible in outside lockers.

In 1932 the first all-steel Magirus Turntable Ladders were delivered to London and Belfast, offering greater rigidity than wooden ladders. Merryweather's first all-steel ladder was delivered to Ilford, Essex in 1933. Not to be outdone, Leyland built a Turntable Ladder for Hull in 1936 with a 150-foot ladder in five sections. A survey in 1936 reported that a total of seventy-eight Turntable Ladders were in service in the UK.

The Spanish Civil War, which commenced in 1936, showed the effects of modern warfare and the incendiary bombing of towns highlighted the future role of the fire service. In 1937 the British government started a programme to supply fire brigades with large numbers of fire pumps.

The Fire Brigades Act of 1938 placed responsibility onto all local councils to provide fire brigades. The new authorities were the county boroughs, non-county boroughs, urban districts and rural districts. There were now 1,440 fire brigades in England and Wales with a further 228 in Scotland.

The ultimate in crew protection was this 1938 forward control Leyland SFT4A limousine Pump Escape acquired by Bournemouth Fire Brigade. It was the brigade's first enclosed appliance and was disposed of in 1960.

At the same time the Auxiliary Fire Service (AFS), a voluntary service, was formed to support brigades in time of war.

By the time of the outbreak of the Second World War, rescue equipment consisted of oxy-acetylene cutting sets and lifting jacks of up to 20-ton capacity with electric tools such as saws and drills powered by on-board generators. Protective clothing included high-voltage gloves, thigh boots and ammonia masks and suits. The rescue of workmen from sewers was a regular occurrence in a fireman's life.

DENNIS BROS

The 1920s had seen Dennis Bros cease using Gwynne pumps and begin using their own pumps based on an Italian design and the Dennis Nos. 1, 2 and 3 pumps emerged. The No. 1 pump was mainly used for trailer pumps while Nos. 2 and 3 were fitted to fire engines. The No. 2 pump was capable of 500gpm and the No. 3 pump 1,000gpm.

In 1928 the Dennis G type was introduced, based on a 30cwt commercial chassis. Pneumatic tyres were fitted as standard (solid tyres were still available as an option) and the appliances were fitted with electric starters, although a starting handle was still fitted and was considered essential up to the 1960s.

The 1930s saw the introduction of a complete new range of Dennis appliances with such descriptive names as Ace, Light Four and Big Six. West Ham Fire Brigade purchased the first Big Six pumping appliance in 1930 followed by another a year later. Both were supplied with Bailey wheeled escapes. Named the Big Six on account of the six-cylinder engine, it was fitted with a mid-mounted pump, which enabled pumping to commence on arrival at a fire without having to remove the escape to gain access to the pump controls. A Big Six with an extended chassis was used for a Turntable Ladder for Belfast in 1932 and fitted with German Magirus steel ladders.

In 1933 Dennis Bros introduced the Ace fire engine and, like all Dennis vehicles, it took its name from the chassis which was normally fitted with either Braidwood or New World bodies. Its short bonnet had a snout-like appearance and it was soon nicknamed the 'Flying Pig'. Small and compact, it was priced at £500 to £1,000 depending on the fittings and equipment required. The Ace was fitted with a four-cylinder engine and by the time the last left the Guildford works in 1939, over 130 had been built.

The Dennis Light Six was introduced in 1935 and fitted with a six-cylinder engine and cost between £790 and £1,275, again depending on the equipment fitted. A total of eighty-four Light Sixes were built with the last leaving the Guildford factory in April 1943. An alternative appliance available was the Light Four fitted with a four-cylinder side-valve engine. It could be supplied with a variety of body styles including enclosed limousine bodies, New World and an unusual style with the officer and driver sitting in an enclosed cab while the crew sat at the rear on Braidwood bodywork. The Dennis Big Four was the heavyweight version of the Light Four, weighing in at nearly 8 tons, and was fitted with an overhead valve engine of 6.8 litres. Again, it was available with the full range of body styles.

LEYLAND MOTORS

The main competitor to Dennis Bros was Leyland Motors but the building of fire engines had never been a large proportion of their business and by 1929 the FE (Fire Engine) range introduced in 1920 was beginning to show its age due to its old designs. However, when the company received an enquiry from Birmingham for a new escape tender a quick decision was made to build one on a Lioness Six coach chassis. It was fitted with a hosereel and delivered by April 1930. Another Lioness Six was delivered in 1931 built with New World bodywork and a side-mounted pump.

Leyland had produced their first Turntable Ladder in 1924, fitted with Metz ladders, and soon built up an excellent reputation. By 1930 the design was old technology but in February of that year an enquiry was received from Nottingham for a revolutionary specification. It was to have Metz 85-foot ladders, a midship-mounted 700gpm pump, a hosereel, and was to be fitted with pneumatic tyres. The appliance was finally delivered to Nottingham in April 1931. The time and effort spent in designing and building this appliance was well spent, as many more orders followed.

The Leyland FK Cub chassis was designed to provide a light fire engine. Two models were available: the FK1 fitted with a Braidwood body and a Reece-Roturbo main pump weighing just over 3 tons and priced at £950; and the FK3, which was designed as a first-aid tender (i.e. without a main pump) and cost £750. By the end of October 1932 the first orders were delivered. In early 1934 two orders were received for Cubs with midship-mounted pumps and these were designated as the FK2 and fitted with transverse seating. When problems arose with the supply of Reece-Roturbo pumps a new supplier was deemed necessary and Gwynne Pumps

This Leyland FK8A semi-limousine Pump, operated by Bristol Police Fire Brigade, was ordered pre-war but not delivered until 1941. It served with the brigade until 1971.

submitted a suitable design. The new model became the FK4 and later upgrades and updated designs became the FK5 to FK9. Nearly 300 Leyland Cubs in various guises had been built by 1941.

Following the success of the FK series, Leyland introduced the FT chassis and the first order for a FT1 was for Rotherham and was completed in February 1932. The FT1 had a rear-mounted pump and when Leyland received an enquiry for one to be built with a midship-mounted pump it was decided to designate it as the FT2. A steady stream of orders followed for both types, including Birmingham, who ordered three FT2s with New World bodywork. The FT3 and FT4 were announced in 1935 and were fitted with a new 700gpm pump and an improved engine.

The Leyland TLM series of Turntable Ladders had been introduced in 1932, fitted with German Metz wooden ladders. By 1933 wooden ladders were being superseded by those made from steel and came in a range of heights from

85 to 164 feet, with two types becoming popular in the UK: the four-section 101-foot and the five-section 104-foot ladder. In total forty-two were delivered by the outbreak of war in 1939, when the supply of Metz ladders ceased.

A new model was introduced by Leyland in early 1938 with a slightly longer chassis than the Cub and was designated the FKT. Of the thirty supplied to civilian brigades, eleven went to London; these could carry either a wheeled escape or an extension ladder and had a fixed monitor, capable of delivering large quantities of water, fitted above the pump. Meanwhile the next design was being developed, which would be the F5 and F7 series. The F5 had a 500–800gpm pump and the F7 a 700–1,000gpm pump. Only twenty-five of this series were built, the last leaving the factory in 1942.

Historically, fire brigades had shunned the use of diesel engines but March 1938 saw another Leyland first, with the delivery to London of a FT3A with a diesel engine. By the time the last FT3A and FT4A models were delivered in 1940, over 150 pumping appliances were built on the FT chassis, including two for the LFB with limousine bodies, which were classed as breathing apparatus pumps.

MERRYWEATHER & SONS

In 1921 the company were advertising a motorised Turntable Ladder with the road engine capable of raising and extending the ladder simultaneously. Rotation of the turntable was still by hand, although by 1922 all operations were mechanised. Also at this time Glasgow Fire Brigade introduced a new rescue apparatus consisting of a lifeline and pulley, allowing casualties to be lowered from the head of the ladder to the ground in a sling. This device was soon to become a standard piece of equipment.

In 1922 the name of Shand Mason disappeared from the fire engineering market when the company was taken over by Merryweather.

THE SECOND WORLD WAR

1943 Austin K4 hand-operated Turntable Ladder fitted with a three-section ladder. Fifty of these were built by Merryweather. This one saw post-war service in West Sussex.

THE SECOND WORLD War created a need for large numbers of pumps and, initially, this was resolved by the production of thousands of trailer pumps which obviously created a requirement of vehicles to tow them. In the early days of the war taxis and other light vehicles were used, while the Auxiliary Towing Vehicle (ATV) was developed. This was built on the Austin K2 chassis with a steel van body providing seating for the crew

and locker space for hose and other equipment. The roof was strengthened to give protection against shrapnel.

Trailer pumps came in a variety of sizes but generally consisted of a two-wheeled unit fitted with its own engine and could be manoeuvred into positions not accessible to motor pumps. Light trailer pumps were intended for rural use with a capacity of up to 175gpm. The large trailer pump with capacities of up to 500gpm was produced in large numbers.

The main self-propelled pump of the war was the Heavy Unit, which consisted of a 700–1,000gpm pump mounted on a lorry chassis. The cab offered seats for the driver and officer with the crew seated in a basic shelter facing to the rear. For convenience of mass production the pumps were powered by a separate engine and not by a power take-off

1942 Austin K4 Heavy Unit registered GLE 801. The Benedictine monks of Belmont Abbey in Herefordshire also served as firemen during the Second World War.

An Escape Carrying Unit of the Second World War. This one was built on a Fordson 7v chassis.

from the road engine. The pumps were manufactured by Gwynne, Sulzer and Tangye and normally had four deliveries. Although there was the occasional use of other chassis, the majority of Heavy Units were built on the Austin K4 and the Fordson 7v chassis.

The outbreak of war had seen an immediate cessation of the import of Magirus and Metz ladders from Germany and in 1940 the Home Office assisted brigades with the acquisition of Merryweather 100-foot steel ladders, which were fitted to the Leyland TD7 bus chassis, the Leyland Beaver chassis or the Dennis Lancet chassis.

The Escape Carrying Unit was a wartime appliance again built on the Austin K4 or Fordson 7v chassis to carry a wheeled escape. It was fitted with a water tank and twin hosereels and an extended tow-bar for pulling a trailer pump. Later models were fitted with Barton front-mounted pumps and thus became Pump Escapes.

The Dam Lorry was an invention of the wartime fire service and was basically a flatbed lorry carrying a dam or water tank. If it carried a portable pump or towed a trailer pump it was known as a Mobile Dam Unit, which was forerunner of the modern Water Tender.

Converted post-war into a Water Tender, this 1943 Fordson 7v Mobile Dam Unit is seen here towing a trailer pump.

In 1943 the Home Office introduced a 60-foot manually operated Turntable Ladder, which cost about a third of the price of a full-size ladder. It was built on the Austin K4 chassis and fitted with a three-section steel Merryweather ladder. Fifty of these were built between 1943 and 1944 and had a standard two-man cab for the driver and officer, with the crew sitting in a rear-facing shelter behind them.

It is only right that any book on the UK Fire Service should include reference to the period of the National Fire Service (NFS). There had been many problems with the mobilisation of back-up crews and appliances from many of the more rural brigades and there was little standardisation of equipment and rank, which made integration difficult; consequently meetings were held between senior officers and the Home Office. The worst air raid on London occurred on the night of 10–11 May 1941, with fires widespread across London. The statistics were frightening with over 500 aircraft bombing the capital – some returning again after refuelling and reloading. It is estimated that 700 tons of high explosive and over 80 tons of incendiaries were dropped in that one night alone, damaging 700 acres of London by fire, with 2,200 fires recorded. Almost 1,500 people were killed, 1,800 badly injured and 12,000 made homeless. Only three days later the Fire Services Bill 1941 was introduced in the House of Commons, receiving royal assent on 22 May and the NFS was formed. It was structured into twelve regions further divided into forty-three fire forces. Standards of uniform, rank markings and training soon followed.

The NFS was never really tested by fire except during the so-called Baedeker raids of 1942 on such historic cities as Canterbury, Exeter and Norwich. Later in the war, however, when Germany subjected the UK to the prolific bombardment of the V1 flying bombs and the V2 rockets in 1944–45, firefighters were heavily involved in rescuing the many trapped casualties.

1945–1959

IN 1948 THE NFS was dissolved and control of fire services returned to local government but on a completely different scale to the pre-war period. The new fire authorities were the cities, county boroughs and county councils and 135 new brigades were formed in England and Wales, eleven in Scotland and four in Northern Ireland.

Replacement of old fire appliances in the immediate period after the end of the Second World War was slow due to a shortage of raw materials being used to replace the country's infrastructure. The Home Office provided a kit comprising a small water tank, pump and hosereel to convert ATVs into Hosereel Tenders, which also towed trailer pumps. Many others

This Dennis F12 Pump Escape was delivered to Somerset in 1952 and is seen here at Bridgwater. These appliances normally had a 1,000gpm pump and carried 100 gallons of water. Some 303 were built for the UK.

were converted into a variety of roles including Emergency Tenders, Hose Layers and Salvage Tenders. Also for an interim period many wartime Heavy Units and Mobile Dam Units were converted into Water Tenders.

The Civil Defence Act 1948 was passed, reconstituting Civil Defence and the AFS with voluntary recruitment. Initially local authorities were allowed the issue of wartime appliances that were being held in store, mainly Heavy Units, ATVs and trailer pumps.

A Joint Committee for the Design and Development of Appliances and Equipment was set up and issued specifications for different types of fire appliance. There were two types of Water Tender specified: Type A, without a built-in pump but towing a trailer pump; and Type B, fitted with a built-in pump. Under the specification both types had to carry 400 gallons of water. Pumping appliances carrying less than 400 gallons were designated as Pumps, or if carrying a wheeled escape as Pump Escapes, but these were mainly in use in the larger towns and cities. It was not until 1949 that the first new appliances began to appear with the Home Office controlling the issue to brigades. The specification for Emergency Tenders distinguished between Type A, which required a minimum crew of six, breathing apparatus, specialist equipment and a built-in generator, and Type B, which was less equipped and a built-in generator was not required, although a portable one was often carried.

Merryweather had designed a 50-foot, three-section alloy ladder and Kent trialled one in 1951 as a replacement for

This Bedford SB Type A Emergency Tender, built in 1953 by Hampshire Car Bodies for Croydon Fire Brigade, carried a full range of rescue equipment and a variety of hand tools.

a wheeled escape. Within two years the brigade decided to replace all its escapes, with the ladders now modified and reduced to 45 feet.

By 1952 designs had been drawn up for new appliances for the AFS. Gradually the new emergency pumps, built on the Bedford SH 4×2 chassis and the RLHZ 4×4 chassis, nicknamed 'Green Goddesses' were issued to replace the wartime vehicles. They used the six-cylinder Bedford petrol engine and were fitted with a 400-gallon water tank and a Sigmund 1,000gpm pump. Later other types of vehicle were issued, including Hose Layers, Pipe Carrying Units, Transportable Water Units (Bikinis) and Control Units. These were built on Bedford SH, Bedford RLHZ or Commer Q4 chassis, while Land Rovers were used as Command Cars.

It was during the 1950s that German Turntable Ladders were re-introduced with Magirus and Metz again making inroads into the UK market. Magirus ladders were fitted onto the Bedford chassis by David Haydon, coachbuilders of Middlesex, and Metz ladders were built onto the Dennis chassis, while those of the only UK manufacturer, Merryweather, were built on the AEC chassis.

One of a pair delivered to Brecon & Radnor Joint Fire Brigade in 1957. The appliance was built on a Bedford J4 chassis with bodywork by Alfred Miles Ltd. Both were Water Tenders carrying 400 gallons of water.

BEDFORD VEHICLES

The SB model fitted with a six-cylinder petrol engine was introduced in 1951, followed in 1953 by the RL, which gave a 4×4 capability, and by the SH in 1954. All these chassis were rated at 10 tons and were capable of the heaviest fire engine duties of the time. Bedford also introduced the A series in 1953, the C and D in 1956 and the J in 1957. These were all 7-ton chassis and were adapted to forward control. All these models became very popular with bodywork built by such companies as Carmichael & Sons, Hampshire Car Bodies (HCB) and Alfred Miles.

Seen at Southsea Fire Station is a 1955 Bedford S Pump Escape. The bodywork was built in Hampshire Fire Brigade's own workshops and was one of many fire appliances that the brigade built.

COMMER TRUCKS

Commer was part of the Rootes Group and its QX chassis was an early post-war model with a six-cylinder engine, weighing in at 6½ tons. It provided a reasonably priced, compact fire engine and was used extensively by rural fire brigades. The 21A and 45A chassis were also used for fire appliances with the 86A chassis being used for a number of 100-foot Magirus Turntable Ladders with David Haydon bodywork.

DENNIS BROS

The F series of fire engines commenced in 1947, starting with the F1 and running through to the F131 some thirty years later, although not all F numbers went into production. The F1 was a development of the pre-war Light Four with an improved engine and was available with either Braidwood or New World bodywork. This was followed by the larger F2 powered by a Rolls-Royce engine with twice as much power as the F1. The F3 was again slightly larger and fitted with

Shown here on driver training duties, this 1952 Commer 21A with Alfred Miles bodywork is working out its last year or so with the brigade. Delivered to Shropshire Fire Service, the appliance ran as a Water Tender Escape in its early years.

a six-cylinder Meadows engine, which enabled the Dennis No. 3 pump to produce 960gpm. The F6 was based on the F1 but with a 600-gallon water tank and was aimed at developing countries where mains water was not readily available. So far all the F series appliances had been aimed at the industrial and export markets.

The Dennis F7, introduced in 1949, used the same eight-cylinder Rolls-Royce engine as the F2; it had a top speed of over 60mph and could accelerate to 40mph in under 30 seconds. The pump was generally midship-mounted, and the vehicle had a very handsome body with a flat front curving in at the lower end. The F8 was a short wheelbase appliance built originally for Northern Ireland. Powered by a six-cylinder Rolls-Royce engine it was narrow and compact and looked like a scaled-down version of the F7. While the majority carried a Dennis body, a large number had bodies by Alfred Miles of Cheltenham. These were totally different in appearance and such appliances carried a portable pump and 400 gallons of water and so were classed as Water Tenders.

The most famous post-war fire appliance ever made, the Dennis F12 was basically the same as the F7 but had a slightly shorter wheelbase. Just about every brigade had at least one and they were sold in vast numbers both at home and overseas, usually as Pump Escapes. Dennis Turntable Ladders of the 1950s used the F14, F17 and F21 chassis, all of which had a 13-foot 6-inch wheelbase and were fitted with the 100-foot Metz ladder, the first of which went to Cambridge in 1953. All of these types still had the F12 style cab.

Dennis Bros had introduced the F24–28 range in the late 1950s and continued to build it for many more years. The F24 (automatic gearbox) and F26 (manual gearbox) were both powered by an eight-cylinder Rolls-Royce engine. The F25 was slightly smaller and was powered by a Dodge six-cylinder engine but was soon replaced by

The Dennis F8 was a short wheelbase appliance powered by a six-cylinder Rolls-Royce engine. In 1953 this one was based at the Leith Fire Station of the South Eastern Fire Brigade in Scotland.

The Karrier-
based fire
appliance was
not a common
sight, there
being fewer than
200 built for
UK brigades.
This Water
Tender built
by Carmichael
of Worcester
was delivered
to Dorset Fire
Brigade in 1961.

the F28 fitted with the Rolls-Royce engine giving some 30 per cent more power. The F27 chassis was used for Turntable Ladders.

KARRIER MOTORS

Another member of the Rootes Group was Karrier Motors whose vehicles were generally smaller than the Commer brand. The Gamecock 72A chassis was used for a small number of fire engines.

LAND ROVER

The first of these was built in 1948, primarily designed for agricultural use but its use as a lightweight fire appliance was soon recognised and hundreds have seen service in fire brigades. As a light pump it had a built-in pump, a water tank, a hosereel and carried a short ladder, hose and other firefighting kit.

LEYLAND MOTORS

In the early 1950s Leyland made a brief foray into the fire engine market using the Leyland Comet chassis with a 5.6-litre diesel engine and stylish bodywork built by such companies as Windovers of Hendon and HCB. The bonneted body, although it kept engine noise from the cab, meant that crew space was reduced in order to keep the overall length down.

In 1958 Leyland returned to the fire engine market with the introduction of the Firemaster chassis, powered by a midship-mounted 9.8-litre diesel engine that allowed the 1,000gpm pump to be front-mounted. Essex had two combined appliances built by David Haydon of Birmingham and designated as Pump Emergency Salvage Tenders. Behind the crew cab was a BA servicing area and a hydraulic rescue kit; cutting equipment and salvage equipment were carried as well as a full range of firefighting equipment and an extension ladder. Two Firemaster chassis were also fitted with Magirus Turntable Ladders but the design was not a commercial success and only eleven Firemasters went into operational service.

The Leyland Firemaster was a new concept in design. It had a front-mounted pump behind twin doors. Sadly the design never caught on and only eleven were produced. This Magirus Turntable Ladder was delivered to Wolverhampton in 1960.

Delivered to Derbyshire in 1956 this AEC Regent carried a Merryweather four-section 100-foot mechanically operated ladder. This type of chassis was normally used for double-decker buses.

MERRYWEATHER & SONS

In the 1950s Merryweather were supplying pumping appliances built on the AEC Regent, Marquis, Mercury chassis and Turntable Ladders on the AEC Mercury and Regal chassis.

THE 1960s AND 1970s

The 1960s and 1970s was a period of great change within the British Fire Service with many amalgamations and the introduction of the term 'Fire and Rescue'. A major amalgamation of brigades occurred in 1965 when the LFB was expanded to include the brigades of Middlesex, Croydon, East Ham and West Ham. Also included were parts of the counties of Essex, Hertfordshire, Kent and Surrey.

Another major re-organisation took place in 1974: county boroughs were abolished with fire brigades now based at county level and England ended up with six metropolitan brigades and thirty-nine county brigades. Wales had eight brigades while Scotland was re-organised in 1975 into eight regional brigades.

BELOW
In 1967 Merryweather produced their Marquis Mk 7 appliance, the design being based on the AEC Mercury TGM chassis. The prototype was delivered to Leicester as a Water Tender Escape but later converted to a Water Tender Ladder.

OPPOSITE
This Control Unit was delivered to Kingston-upon-Hull in 1959. It consisted of a Maudslay Merlin 2 chassis with Merryweather bodywork. The use of this chassis for fire appliances in the UK was unusual, with only forty delivered.

Although fire brigades did not have a statutory duty to respond to non-fire emergencies they had always been equipped to do so and in 1974 an innovative move by the Chief Officer of Cambridgeshire, John Maxwell, changed the name of the brigade to Cambridgeshire Fire and Rescue Service. A few others followed but nearly thirty years were to go by before the Fire Brigades Act of 2004 gave brigades the statutory right to be the rescue service of the UK. Nearly all brigades are now termed 'Fire and Rescue Service'.

The first Hydraulic Platform was built in 1963 by Simon Engineering, with a working height of 65 feet. By the late 1960s they were being built up to 85 feet high. In 1972 the company introduced a new range with working heights of up to 103 feet, and these were marketed alongside the 85-foot models. However, by 1975 the smaller model had been discontinued. Pump Hydraulic Platforms were a short-lived attempt at building a replacement appliance for Pump Escapes. They were basically a pumping appliance fitted with a 50-foot hydraulic platform.

The mid-1960s saw the introduction of the Lacon ladder, which would eventually replace all wheeled escapes in the UK fire service. Originally known as the '464', it was 46 feet

The first Hydraulic Platform to go into service in the UK was delivered to Monmouthshire in 1963. This was a Commer VA fitted with 65-foot Simon booms.

4 inches long when extended and became the primary rescue ladder for fire brigades, but was later re-designated as the '135' (13.5 metres). To differentiate between appliances, those carrying the new ladder were classed as Water Tender Ladders (WrLs), while those carrying a 30-foot or 35-foot ladder were classed as Water Tenders (WrTs); for the purposes of this book all appliances have been called WrTs as the roles are interchangeable.

In 1967 Glasgow trialled a detachable cage that could be fitted to the head of a Magirus Turntable Ladder and was soon fitted to all of its Magirus ladders. Eventually all new Turntable Ladders were fitted with cages.

Bedfordshire introduced the first air-lift bags in 1969, which were inflated from breathing apparatus cylinders. The German company Lukas designed the first hydraulic rescue tool in 1972, capable of cutting through and spreading impacted metal; these proved invaluable and are now carried by nearly every front-line fire appliance in the UK.

The 1970s saw an influx of new manufacturers with Anglo Coachbuilders (later renamed Angloco) and Eagle Engineering of Warwick building new appliances.

With the advent of motorways the need for faster and lighter rescue tenders became necessary. Land Rovers had always been popular but in 1972 Carmichael converted the Range Rover model by adding an extra trailing axle giving a 6×4 configuration. This enabled a greater payload to be carried including heavy rescue equipment.

BEDFORD VEHICLES

The TK range of chassis had been introduced in 1960 and soon became a favourite with fire brigades. This range, which included the TK, TKL, TKEL and TKG series, was used for hundreds of fire engines. Although Bedford only supplied the chassis, many bodywork builders including Carmichael and HCB all constructed Bedford-based fire engines.

Seen here is one of two Bedford TK Water Tenders built for Wiltshire Fire Brigade by Mumfords in 1963.

HCB-Angus built its first CSV (crew safety vehicle) in 1976 on the Bedford TKG chassis. It was promoted as being able to withstand greater crush and impact loads than any other known design. An alternative was the high-strength cab, reinforced with steel tubing giving improved crash resistance without going to the levels of the CSV.

COMMER TRUCKS

In 1963 the VAC chassis was introduced and was used for pumping appliances, with Kent ordering a batch of twelve. Commer stopped production of a chassis suitable for fire appliance use in the mid-1970s.

DENNIS BROS

The D type introduced in 1967 was designed to replace the F8, being a small appliance suitable for both rural districts and congested city centres.

By the 1970s Dennis's bodyshop had changed its techniques by using glass fibre-reinforced plastic instead of steel and aluminium panels. The F36 (eight-cylinder Rolls-Royce engine) and the F38 (six-cylinder Rolls-Royce engine) were both fitted with new moulded front panels. For many years Dennis had not considered using diesel engines but with the introduction of a suitable Rolls-Royce engine the first diesel-engined appliance was built by Dennis and for the next twenty years diesel-powered appliances would be built alongside petrol-engined counterparts but with F numbers from F101 upwards.

Delivered in 1968 to the Northern Area Fire Brigade, this Commer VBKW Water Tender with HCB-Angus bodywork is seen here at the retained fire station in Kingussie.

The Dennis R series was introduced in 1976 by the company now renamed Hestair Dennis, with the ash-framed cab replaced with a tubular steel structure integrated with a glass fibre moulding. The R series was available with diesel or petrol engines, automatic or manual gearbox, multi- or single-stage pumps in gunmetal or alloy. By the late 1970s the use of petrol engines in Dennis fire engines was coming to a close with the last of the F series the F60 and F61.

This Dennis F48 Water Tender Ladder fitted with a Rolls-Royce engine was delivered to the Peterborough Volunteer Fire Brigade in 1971. The brigade, the only one of its kind in the UK, has restored it and it is kept for ceremonial purposes.

In 1978, Dennis introduced its most successful fire engine, the RS, of which over 1,750 were built. This was a continuation of the R series but with a new steel cab.

A one-off appliance was this Dodge K1113 Water Tender Escape with bodywork by Jennings for Bristol Fire Brigade in 1972.

DODGE VEHICLES

The first K850 appliance appeared in 1970, fitted with a Perkins diesel engine. By the late 1970s the G13 series was beginning to replace the K series. The G13 had a more powerful Perkins V8 diesel engine and a six-speed gearbox. The heavier Dodge G16 chassis was available for mounting Turntable Ladders and Hydraulic Platforms.

ERF

An established manufacturer of heavy vehicles, which in 1966 launched two new chassis for the fire engine market. The 84RS model was designed for WrTs while the heavier 84PF was used for Turntable Ladders and Hydraulic Platforms.

Fire appliances on the ERF 84RS chassis first appeared in 1966. This is one of three Water Tenders delivered to Staffordshire in 1968, built by HCB-Angus.

FORD MOTORS

In 1965 Ford launched the D series 4- to 10-ton chassis with a 6-litre diesel engine. Over the next ten years many WrTs were built on the Ford D1014 chassis with Emergency Tenders, Hose Layers and other special appliances using the heavier D1317 and D1617 series. Also for many years the Ford Transit series was used for light pumps and Rescue Tenders.

LAND ROVER

The 109 chassis was introduced in 1968 and companies such as Carmichael built conversions as light pumps.

SHELVOKE & DREWRY

In 1975 it entered the fire engine market in conjunction with Carmichael & Sons, building WrTs on the WX chassis and aerial appliances on the heavier WY chassis.

This 1972 Ford D800 Pump Escape was built by Pyrene for Gloucester City Fire Brigade as a one-off. Note the different locker doors and the position of the pump controls to allow it to be operated without removing the escape ladder.

The Land Rover chassis was used in considerable numbers for fire engines. This FT6 was built in forward control form by Carmichael and was one of a pair supplied to Shropshire as Rescue Tenders.

THE 1980s AND 1990s

B Y THE 1980s and 1990s there were many changes in the fire services' fleets with the withdrawal from the fire engine market of many famous names and the introduction of foreign manufacturers.

The decline of the British truck industry had a major effect on fire engine manufacturers by reducing the variety of chassis available. Dennis and Dodge, now under the Renault name, remained but ERF, Ford and Shelvoke & Drewry left the fire engine market in the early 1980s, followed by the closure of Bedford in 1988. They were replaced by the return of Leyland and from mainland Europe by MAN, Mercedes-Benz, Scania and Volvo.

This Dennis Sabre TSD233 Water Tender Ladder was one of fourteen purchased by Devon between 1997 and 1998, all built by John Dennis Coachbuilders.

Cheshire Fire Engineering, a subsidiary of ERF, closed in 1982 but Saxon Sanbec, founded by former Cheshire Fire Engineering staff, rose from the ashes and continued building fire engines. Mountain Range of Crewe had been exporting fire engines for a few years and with the opening of a new factory in 1982 entered the UK market but then ceased trading in 1991.

Merryweather, having moved from London to South Wales, suddenly moved to Plymouth in 1984 and was closed down by parent company Siebe Gorman. The demise of Merryweather left the UK without a manufacturer of Turntable Ladders and brigades had to rely on imports of Magirus and Metz from Germany and Camiva from France.

Hestair Dennis had been re-organised in 1985 with Dennis Specialist Vehicles formed to produce the chassis for fire engines. John Dennis Coachbuilders (JDC) was formed as an independent company by former Hestair Dennis employees and over the intervening years JDC has built a reputation for quality. In addition to building traditional WrTs on Dennis, MAN, Mercedes-Benz, Scania and Volvo chassis they also build specialist appliances on other chassis including Land Rover, Pinzgauer, Steyr Puch and Unimog.

In 1986 there was a new entrant to the fire engine market, Excalibur CBK, who built a range of fire engines for brigades across the UK using various chassis including Dennis, Mercedes-Benz and Volvo. Another new entrant to the fire engine market in 1990 was Halton Fire Engineering who supplied Mercedes-Benz based appliances.

After Bedford left the market many brigades soon had to make a decision on a replacement chassis and by 1990 Volvo were claiming 40 per cent of the market.

1991 saw Fulton & Wylie cease trading and Emergency One (UK) was formed by former staff and is now one of the leading UK builders of fire appliances. HCB-Angus was another company to cease trading in 1991.

The Dutch company Plastisol had been making injected resin bodies for airfield fire tenders for a number of years when The Vehicle Application Centre (TVAC), of Leyland in Lancashire, approached them in 1996 to manufacture a body for a pumping appliance. The body and water tank were produced as one unit, which offered a significant weight saving. Lancashire purchased the first such appliance, built on a Leyland-DAF chassis.

The last wheeled escape in the UK was withdrawn from service in 1994, which was the end of an era going back to Victorian times. The late 1990s saw the first Aerial Ladder Platform in use in the UK. It offered a versatility not seen before, being able to go up and over roofs. The main boom was a telescopic two-section unit with a cage attached to a small end boom with a working height of 29.5 metres and an outreach of 18 metres and the booms are fitted with a telescopic rescue ladder. Later models are being built with a working height of 42 metres.

In 1997 Magirus introduced an articulated Turntable Ladder in which the top 3.5-metre section articulated below the horizontal, similar to the Aerial Ladder Platform. Cambridgeshire purchased two of these ladders built on the MAN 18.264 chassis.

The 1990s saw Rescue Pumps appear in fire service fleets for the first time. These are externally identical to WrTs but in addition to their firefighting equipment they carry enhanced rescue equipment, often including water rescue and line rescue equipment.

The Local Government Act (Wales) of 1996 saw a further reduction of the number of Welsh fire brigades to three based on regions.

DENNIS BROS

The DS model was introduced as a replacement for the D series and the first production model went into service in

John Dennis Coachbuilders built nineteen Water Tender Ladders for Cambridgeshire Fire and Rescue Service on the Dennis RS237 chassis between 1989 and 1993.

1981. The larger SS model, similar to the RS series but with a tilting cab, came in 1982 when the LFB ordered forty.

The first Dennis Rapier appliance was delivered in 1991 powered by a Cummins six-cylinder turbo-charged diesel engine and with its independent front suspension and space frame chassis was probably the most advanced fire appliance ever built.

In an effort to reduce costs a consortium of brigades produced a specification for a standard WrT, to succeed the Dennis RS and SS models. Equipment stowage needed to be flexible for individual brigades and sliding trays and mechanisms were used for easy access in particular for heavy equipment. The final design was the Dennis Sabre built by JDC on a conventional ladder frame chassis and the first delivery went to Wiltshire in 1995. There were a number of options available including the XL cab capable of accommodating a crew of

Devon Fire and Rescue Service purchased three Dennis Rapier TF203 Water Tender Ladders in 1994, all built by John Dennis Coachbuilders.

ten and a heavy-duty chassis designed for Turntable Ladder use.

ERF

The company had ceased building fire engines in 1982 but re-emerged in 1996 when a number of appliances were built on the heavy-duty EC8 and EC10 chassis. These included Foam Tenders for Cheshire and Greater Manchester and Aerial Ladder Platforms for West Midlands.

This 1988 Leyland 180 Water Tender Ladder was built for Lancashire Fire and Rescue Service by Fulton & Wylie.

LEYLAND DAF

In 1987 DAF had acquired Leyland Trucks and since then DAF-badged fire engines have been in use in the UK. One brigade, Lancashire, has ordered over seventy DAF appliances.

1995 saw a large delivery to Devon Fire and Rescue Service when Saxon built nineteen of these MAN 10-223F compact Water Tender Ladders.

MAN TRUCKS

The first order came from Devon in 1994 with an order for twenty-five pumping appliances on the L2000 chassis, all built by Saxon Sanbec and in 1999 Gloucestershire ordered a number of compact WrTs for use in rural areas which were built by JDC on the MAN 10.224 chassis.

MERCEDES-BENZ

Mercedes-Benz firefighting appliances appeared on the scene with the introduction of the 1120AF chassis in 1992, which was followed in 1994 by the 1124F; the 1625, 1726 and 1827 series were used for aerial appliances.

SCANIA

The first Scania fire engines in use in the UK were Turntable Ladders in the early 1980s followed by WrTs built on the G92M chassis. Later models were built on the P94D chassis powered by a DSC9 diesel engine.

This 1997 Mercedes-Benz 1124F Water Tender Ladder was built by Saxon for Somerset Fire and Rescue Service and was part of an order of fourteen delivered between 1992 and 1997.

VOLVO TRUCKS

The mid-1980s saw the first appearance of Volvo fire engines in the UK with the introduction of the FL6.14 chassis. The first large order came from London with an order for thirty-four pumping appliances and six Aerial Ladder Platforms on the FL10 8×4 chassis.

In 1992 the Volvo FL6.18 chassis appeared along with the higher-powered FS7. Five of the smallest water tenders ever built by Volvo for the UK market, the FL6.11, were delivered to Cumbria for use in the narrow and winding lanes of the Lake District.

THE TWENTY-FIRST CENTURY

Modern fire appliances are technologically more advanced compared to the steam-powered pumps of the nineteenth century and carry a variety of equipment and firefighting media to deal with fires, rescues, vehicle extrication, floods and salvage. Whereas in the past fire engines were built as complete units by a manufacturer such as Dennis Bros, Leyland Motors and Merryweather & Sons, the building of modern appliances involves three separate processes. Firstly, the chassis and cab are supplied by a commercial truck manufacturer, such as DAF, Iveco, MAN, Mercedes-Benz, Scania or Volvo. Secondly, a polypropylene modular body is moulded by a specialist company. Finally the two components are married together and fitted out by fire

Volvo FL260 Enhanced Response Pump built by Emergency One UK – one of eighteen in service with Hampshire Fire and Rescue in 2017. This is a standard Rescue Pump carrying both 13.5m and 9m ladders and a full range of firefighting and rescue equipment.

appliance manufacturers. These companies include Angloco, W.H. Bence, Emergency One UK and JDC. Light vans such as the Iveco Daily and Mercedes-Benz Sprinter are also converted and in use as Command Support Units, Rescue Tenders and other support roles.

When Saxon designed the new Volumax cab, Somerset Fire and Rescue Service ordered twelve Water Tender Ladders built on the Volvo FL6.14 chassis for delivery in 2000.

One of the MAN TG-A 26.363 Incident Response Units built by Marshall SV, seen here with its Moffett Mounty M2003E forklift truck.

A MAN TG-A 26.363 6×4 Prime Mover seen here unloading an USAR pod. The Partec Multilift system can clearly be seen.

The latest firefighting pumps are electronically controlled. At the touch of a button the pump is engaged, the tank valve opens, water flows into the pump and the hosereels are pressurised, ready for instant use. The pump is controlled by pre-set pressures, which can also be increased or decreased at the touch of a button.

Saxon designed a new cab for the Volvo FL6 Water Tenders in 2000 named the Volumax, which gave more headroom and additional storage area but in 2004 ceased trading. Excalibur also disappeared from the UK market in 2004, followed by the most famous name in British fire engineering history in 2007 when Dennis finally ceased production of a fire engine chassis. TVAC followed suit in 2008 and also ceased business.

Following the terrible events on 11 September 2001 in America the UK's response was to produce a specialised fleet of vehicles under the New Dimensions

Kuiken Hytrans High-Volume Pumping Unit, seen here at work at a large fire in Thetford Forest.

programme to deal with terrorist attacks, major flooding and other disasters. The first vehicle in the fleet is the Incident Response Unit built on the MAN TG-A

Urban Search and Rescue Unit Module 1. With the side access doors open the method of storage can clearly be seen.

26.363 6×2 chassis by Marshall SV of Cambridge, which carries two decontamination units, each capable of treating two hundred casualties per hour. The units are tent-like structures of three sections for undressing, showering and re-dressing. The equipment is carried in container pallets and unloaded by a forklift truck, which is carried on the rear of the vehicle.

Marshall was then awarded a contract to supply over two hundred Prime Movers, this time built on the 6×4 version of the same MAN chassis. The Prime Movers are fitted with a Multilift hook-lift installation with a capacity of 20 tons and are used for carrying demountable pod units including high-volume pumping units (HVPU), urban search and rescue units (USAR) and mass decontamination disrobe and re-robe Units.

The HVPUs were supplied by Kuiken Hytrans of Holland and each set consists of one pump unit, three hose boxes, and other equipment. A Prime Mover carries either a pump unit and one hose box or two hose boxes. The complete unit is capable of pumping large volumes of water over distances of nearly 2 miles.

Hampshire's First Response Pump is based on an Iveco 70-210 chassis built by Emergency One UK. It is fitted with a 1,000-litre water tank, two breathing apparatus sets, 7-metre ladder, battery-operated cutting and spreading tool and other firefighting equipment.

USARs consist of five modules in container-like pod units supplied by Bootle Containers and fitted out by JDC. Module 1 carries equipment for scene safety, technical search, lighting and timber and concrete cutting and drilling. Module 2 contains heavy cutting, lifting and confined space and rope access equipment for use in major transport incidents. Module 3 supports Module 1 at structural collapses and includes timber supports, heavy breaking and breaching tools, heavy lifting equipment, access platforms and lighting. Module 4 is a drop-side unit carrying a logistics and servicing structure and a Bobcat multiple-purpose vehicle capable of transporting equipment and moving rubble to clear sites. Module 5 is a flatbed unit carrying 10 tons of pre-cut timber for supporting unstable structures.

The mass decontamination disrobe and re-robe units are similar to the USAR pods and are designed to replenish the IRUs as decontamination progresses. The disrobe unit contains 1,600 undressing packs and the re-robe unit contains 1,500 packs for decontaminated casualties.

Volvo Intermediate Response Pump built on a Volvo FL260 12-ton chassis by Emergency One UK. It is fitted with a 1,300-litre water tank, one hosereel, Cobra Cold Cut firefighting system, three BA sets, ladder, battery-powered cutting equipment including spreaders and rams, PPV fan, portable pump and other firefighting equipment.

The final vehicle in the programme is the Detection, Identification and Monitoring Unit based on the Iveco Daily 50C17 van. The equipment carried allows the team to detect and analyse a range of chemicals, radiological materials and biological hazards in various states. Gas-tight suits and breathing apparatus are carried for the team's protection.

In 2012 the eight regional fire brigades in Scotland were amalgamated into one: the Scottish Fire and Rescue Service. In England two joint services were created when Devon and Somerset Fire and Rescue Services amalgamated in 2007 and Dorset and Wiltshire joined forces in 2016.

A number of Fire and Rescue Services are investigating the use of smaller, lighter fire appliances and in 2013 West Midlands introduced Brigade Response Vehicles based on the Toyota HiLux 4×4. These vehicles are used for a quick response and attend such incidents as rubbish and chimney fires but are also used as backup to standard WrTs. Devon & Somerset came up with a different solution in 2014 by purchasing Light Rescue Pumps built on the Iveco Eurocargo chassis. Hampshire is carrying out trials in the use of three levels of pumping appliance: the First Response Vehicle based on the Iveco 70-210; the Intermediate Response Pumps built on the Volvo FL260 12-ton chassis; while the Enhanced Response Pump is a standard Rescue Pump based on the Volvo FL260 15-ton chassis.

DAF

In 2016 Lancashire completed a replacement programme with Emergency One supplying DAF LF250 appliances built on a 16-ton chassis.

DENNIS BROS

In 2000, following a demand for a compact fire engine built at a lower cost, Dennis attempted to design a cheaper version of the Sabre named the Dagger. Although it was fitted with

One of the last Dennis appliances ever built was this Dennis Sabre Water Tender Ladder delivered in 2006 by John Dennis Coachbuilders to Gloucestershire.

a less powerful engine and with some of the non-essential features of the Sabre removed, the cost saving was minimal and only a few were built.

MAN TRUCKS

Avon Fire Brigade put the first of twenty new pumps on the run in 2000, built by Saxon; another large order for the MAN series came from Devon, which had forty-three delivered by 2006.

In 2015 Gloucestershire Fire and Rescue Service purchased five MAN TGL 12.250 Water Tender Ladders, built by Emergency One.

One of twenty-four Mercedes-Benz Atego 1328 Water Tender Ladders on the run in Cornwall, in this case built by Carmichael & Sons.

MERCEDES-BENZ

When the Atego chassis was launched in 2000, among the large orders were one from Cornwall for twenty-four appliances and one from London who replaced their complete front-line pumping fleet of two hundred vehicles over a nine-year period from 2002.

SCANIA TRUCKS

The P270 series was introduced in 2004 and Scania now supply appliances to nearly half of the UK's fire brigades.

Four Scania P274 Water Tender Ladders were delivered to Gloucestershire Fire and Rescue Service between 2010 and 2011 by Emergency One.

AIRPORTS AND INDUSTRIAL BRIGADES

AIRPORTS

Before the Second World War, airfield fire protection was minimal with appliances usually consisting of converted vehicles. It was not until after the war that the development of commercial aviation and the growth in the size of passenger aircraft created the need for adequate fire protection. Airport fire and rescue services are equipped with appliances that have a cross-country capability, with an ideal response time of two minutes but not exceeding three minutes to any part of the airfield. Consequently airport fire stations are usually located close to runways and taxiways to give quick access for their appliances. The appliances are fitted with roof- and bumper-mounted monitors, which deliver large quantities of firefighting foam.

Carmichael Cobra 2 6×6 foam tender delivered to Cambridge Airport in 2009. It carries 10,000 litres of water, 1,200 litres of foam and 100 kilograms of dry powder.

This Bedford RL Water/Foam Tender built by HCB-Angus was delivered to the College of Air Training, Hamble Airfield in 1968. It carried 400 gallons of water and 40 gallons of foam.

INDUSTRIAL

In the nineteenth and early twentieth centuries many manufacturing plants operated their own fire brigades due to the lack of well-organised brigades even in relatively large towns. Regrettably, the number of brigades has diminished over the years, some by disbandment and many by site closure, although the value of maintaining onsite brigades for a relatively small sum can often avoid costly fires and loss of income. Many works brigades rely on the company's ordinary

1975 Range Rover 6×4 Rapid Intervention Vehicle supplied to Birmingham Airport by Carmichael. It carried 200 gallons of water and 20 gallons of foam.

This Water/ Foam Tender was delivered new to the Dartford factory of the Wellcome Foundation in 1988. It was built on a Ford Cargo chassis by HCB-Angus.

workforce to provide a part-time brigade who quickly leave their workplace, don their uniforms and rush to the incident. Some of the larger industries have built fire station complexes crewed by a team of full-time firefighters who have been trained to deal with the particular dangers caused by the site's operations and products. Many of the brigades operate second-hand local authority appliances, while petrol refineries and chemical plants have specially designed vehicles relevant to their individual operations.

This Volvo FL6-17 Dry Powder/Foam Tender was built in 1992 by Reynolds-Boughton for the Coryton Refinery in Essex. Note the twin roof monitors for delivering the firefighting media.

GLOSSARY

Aerial Ladder Platform (ALP): An appliance with folding, telescopic booms that rotate through 360 degrees, and fitted with a cage, monitor and ladder.

Command Support Unit (CSU) or Control Unit (CU): A mobile control room containing computers, maps, radios and satellite communications.

Emergency Tender (ET): An appliance carrying a large variety of tools and rescue equipment; basically a mobile workshop.

Hook Ladder: A wooden ladder 13 feet long with a hooked, folding bill with serrations on the underside. The ladders hooked onto window ledges and hung vertically.

Hose Layer (HL): A fire engine capable of laying out hose over long distances. Lengths of hose are carried already coupled together.

Hydraulic Platform (HP): An appliance fitted with a hinged boom, of two or three sections, capable of rotating through 360 degrees, with a cage at the top.

Pod: A box-type unit built to be carried on a Prime Mover.

Prime Mover (PM): An appliance consisting of a crew cab and an open chassis onto which a pod can be loaded and unloaded.

Pump (P): A pumping appliance carrying a 30-foot ladder and a limited water capacity. Also a generic term for any fire engine carrying a crew of four to six, breathing apparatus, ladders and other firefighting equipment and capable of pumping water.

Pump Escape (PE): Pumping appliance similar to a pump but carrying a 50-foot wheeled escape.

Rescue Pump (RP): A Water Tender Ladder carrying enhanced rescue equipment.

Rescue Tender (RT) or Rescue Vehicle (RV): An appliance carrying a large variety of tools and rescue equipment.

Turntable Ladder (TL): Appliance fitted with a 100-foot extending ladder capable of rotating through 360 degrees.

Water Tender (WrT): A pumping appliance carrying a 35-foot ladder and 400 gallons of water.

Water Tender Ladder (WrL): Same as a Water Tender but carrying a 45-foot ladder.

PLACES TO VISIT

The following list is far from exhaustive but features some museums that display historic fire engines and other memorabilia. It is advisable to check their websites for opening hours.

Banwell Fire Station Museum, East Street, Banwell, Somerset, BS29 6BN. Website: www.banwellfire.weebly.com

Essex Fire Museum, Grays Fire Station, Hogg Lane, Grays, Essex, RM17 5QS.
Website: www.essex-fire.gov.uk/Fire_Museum

Greater Manchester Fire Service Museum, Maclure Road, Rochdale, OL11 1DN.
Website: www.gmfsmuseum.org.uk

Greenock Museum & Heritage Centre, The Old Fire Station, Dalrymple Street, Greenock, PA15 1LY.
Website: www.sfrheritagetrust.org

Hertfordshire Fire Brigade Museum, Lower High Street, Watford, WD17 2AG.
Website: fire.watfordmuseum.org.uk

Kent Firefighting Museum, Woodlands Garden Centre, Ash Lane, Ash, Kent, TN15 7EG.
Website: www.kentfirefightingmuseum.org.uk

London Fire Brigade Museum, The Workshop, Lambeth High Street, London SE1 7JS.
Website: www.london-fire.gov.uk

National Emergency Services Museum, The Old Fire Station, West Bar, Sheffield, S3 8PT.
Website: www.emergencymuseum.org.uk

Stalham Firehouse Museum, High Street, Stalham, Norfolk, NR12 9BB.
Website: www.stalhamfirehousemuseum.co.uk

Welsh Museum of Fire [Welsh Area Fire Engine Restoration Society], Unit 31, Lonla Village Workshops, Skewen, Neath, SA10 6RP.
Website: www.facebook.com/Wales.Fire.Museum

FURTHER READING

Baker, Eddie. *A History of Firefighting in Cambridgeshire*. Jeremy Mills Publishing, 2006.

Baker, Eddie. *Airfield Firefighting in Eastern England*. Jeremy Mills Publishing, 2010.

Baker, Eddie. *Fire Appliances of Eastern England*. Jeremy Mills Publishing, 2007.

Baker, Eddie. *Fire Appliances of South West England*. EB Books, 2015.

Baker, Eddie. *Industrial Firefighting in Eastern England*. Jeremy Mills Publishing, 2014.

Baker, Eddie. *On The Run: A History of Croydon Fire Brigade*. Jeremy Mills Publishing, 2004

Bunn, Mike. *Fighting Fire in Essex: An Illustrated History*. EB Books, 2016.

Bunn, Mike. *Fighting Fire in Warwickshire: An Illustrated History*. EB Books, 2018.

Henderson, Ronald. *British Steam Fire Engines*. Amberley Publishing, 2016.

Hickin, William. *Capital Brigade*. WFH Publications, 2014.

Hickin, William. *Fire Force*. WFH Publications, 2013.

Hutchinson, Barry. *Dennis Fire Engines*. Amberley Publishing, 2015.

INDEX

Page numbers in bold refer to illustrations

AEC 28, **34**: Marquis 34; Mercury 34;
Merlin 34; Regal 34; Regent 34
Aerial Ladder Platform 44, 46
Air bags 37
Airports 57–8
Albion 12, 15
Angloco 37, 49
Austin 22: K2 22; K4 22, **23**, 24, 25
Auxiliary Fire Service (AFS) 17, 27, 28:
Command Car 28; Control Unit 28;
Green Goddess 28; Hose Layer 28; Pipe
Carrying Unit 28; Transportable Water
Unit (Bikini) 28
Auxiliary Towing Vehicle (ATV) 22, 27
Barton pumps 24
Bedford Motors 2, 28, 37–8, 42: A series
29; J4 **28**; RL **58**; SB **27**, 29; SH 28,
29; RLHZ 28; TK 2, 37, **38**; TKG 38
Bence, W.H. 49
Braidwood, James 6, 13, 19, 29
Braithwaite, John 6, 18
Brigade Response Vehicle 54
Cambridgeshire 36, 44: Fire and Rescue
Service 36
Camiva 43
Carmichael & Sons 29, 37, 41, 56:
Cobra **57**
Cedes 11, 13
Cheshire Fire Engineering 43
Civil Defence 27
Command Support Unit 49
Commer 12: QX 29; Q4 28; 21A **30**; VA
36; VAC 38; VBKW **39**
Corps of Vigiles 4
Croydon 35: Fire Brigade **27**; Fire Station **12**
DAF 44, 46, 48, 54: LF250 54
Dam Lorry 24
Dennis Bros 2, 10, 13, 15, 17–19, 24, 31,
38–40, 42, 43, 44–6, 48, 50, 54–5: Ace
15, 18; Big Six 18; Big Four 18; Dagger
54; DS model 44; D type 38, 44; F7 30,
31; F8 **31**; F12 **26**; 31; F14 31; F15 2;
F17 31; F21 31; F24 31; F25 31; F26 31;
F27 32; F28 32; F36 38; F38 38; F48 **39**;
F60 39; F61 39; F101 38; F series 2, 29;
G type 17; Hestair Dennis 39, 43; Light
Four 18, 29; Light Six 18; N type 12;
Rapier 45, **45**; R series 39, 40; RS 40, 45,
45; Sabre 42, 55, **55**; SS model 45
Detection, Identification and Monitoring
Unit 54
Devon and Somerset Fire and Rescue
Service 54
Dodge 31, 40, 42; G13 40; G16 40;
K1113 **40**; K850 40
Dorset and Wiltshire Fire and Rescue
Service 54
Eagle Engineering 37
Emergency One (UK) 43, 49, 54
Emergency Tender 10, 27, 41
Enhanced Response Pump **48**, 54
ERF 40, 42, 46; EC8 chassis 46; EC10
chassis 46, 84PF 40; 84RS 40, **40**

Escape Carrying Unit 23, 24
Escape Vans 8
Excalibur CBK 43, 50
Fire and Rescue Service 36
Fire Brigades Act 1938 16
Fire Brigades Act 2004 36
Fire Services Bill 1941 25
First Response Pump **52**, 54
Foam Tender 46
Ford Motors 41, 42: Cargo **59**; D series
41; D1014 41; D1317 41; D1617 41;
D800 **41**; Transit 41
Fordson 7v **23**, 24, **24**
Fulton & Wylie **43**
Glasgow 37; Fire Brigade 21
Great Fire of London 5
Gwynne pump 12, 17, 19, 24
Halton Fire Engineering 43
Hampshire Car Bodies (HCB) 29, 33,
37, 38, 43
Hampshire Fire Brigade 2
Haydon, David 28, 29, 33
Heavy Unit 23, 27
High Volume Pump (HVPU) **51**, 52
Hook ladders 8
Hose Carts 7
Hose Layer 27, 41
Hosereel Cart 7
Hydraulic Platform 36, 40
Incident Response Unit (IRU) **49**, 51, 53
Intermediate Response Pump **53**, 54
Iveco 48: Daily 49; Eurocargo 54; 70-
210 54
John Dennis Coachbuilders (JDC) 43, 45,
45, 47, 49, 53
Karrier **32**: Gamecock 72A 32
Kuiken Hytrans 51, 52
Lacon ladder 36
Land Rover 28, 32, 37, 41, 43: FT6 41;
109 chassis 41
Leyland Motors 12, **14**, **15**, **16**, **17**,
19–21, **20**, 24, 42, 46, **46**, 48: Beaver
24; Comet 33; Cub 19; FE range 19;
Firemaster 33, **33**; FK 19; FT 20; FKT
21; Lioness Six 19; TD7 24; TLM 20
Light Rescue Pump 54
Local Government Act (Wales) 1996 44
London Fire Brigade (LFB) 6, 35, 45, 56
London Fire Engine Establishment (LFEE)
5, 6
Lukas 37
Magirus 16, 18, 24, 28, 29, 33, 37, 43, 44
MAN Trucks 42, 43, 47, 48, 55: L2000
chassis 47; 10.223F chassis 46; 10.224
chassis 47; 12.250 chassis 55; 18.264
chassis 44; 26.363 chassis 51
Manchester Fire Brigade 10
Manual pump 2, 4, 5
Marshall SV 52
Mass Decontamination 52
Maxwell, John 36
Mercedes-Benz 42, 43, 48, 56: Atego
56, **56**; Sprinter 49; 1120AF chassis
47; 1124 chassis 47; 1625 series 47;
1726 series 47; 1827 series 47

Merryweather & Sons **5**, 12, 15, 16, 21,
24, 25, 27, 34, **35**, 43, 48
Metropolitan Fire Brigade (MFB) 6, 7
Metz 19, 20, 21, 24, 28, 31, 43
Miles, Alfred 29, 30
Mobile Dam Unit 24, 27
Morris, John 10
Mountain Range 43
Multilift 52
National Fire Service (NFS) 25
New Dimensions 50
New World 14, 18, 19, 29
Newsham, Richard 4, **4**, 5
Peterborough Volunteer Fire Brigade **39**
Pinzgauer 43
Plastisol 44
Prime Mover **50**, 52
Pump 11, 27
Pump Emergency Salvage Tender 33
Pump Escape 24, 27, 31, 36
Pump Hydraulic Platform 36
Range Rover 37, **58**
Reece-Roturbo pump 19
Renault 42
Rescue Pump 44
Rescue Tender 41, 49
Rolls-Royce 29, 30, 32, 38
Salvage Tender 27
Saxon Sanbec 43, 47, 55: Volumax cab
49, 50
Scania Trucks 42, 43, 47, 48, 56: G92M
chassis 47; P270 series 56; P274 **56**;
P94D chassis 47
Scotland 35
Scottish Fire and Rescue Service 54
Shand Mason 6, 21
Shelvoke & Drewry 41, 42
Sigmund pump 28
Simon Engineering 36
Squirts 5
Steam fire engine 6, 10
Steyr Puch 43
Sulzer pump 24
Tangye pump 24
The Vehicle Application Centre (TVAC)
44, 50
Tilley manual fire engine 2
Tilling-Stevens 13
Toyota HiLux 54
Turntable Ladders 8, 13, 16, 19, 20, 25,
28, 29, 31, 37, 40, 43, 44, 46, 47
Unimog 43
Urban Search and Rescue (USAR) **51**,
52, 53
V1 flying bomb 25
V2 rocket 25
Volvo Trucks 42, 43, 47, 48: FL260 48,
54; FL6.11 47; FL6.14 47; FL6.17 **59**;
FL6.18 47; FL10 47
Wales 35
Water Tender 24, 27, 30, 37
Water Tender Ladder 37
West Ham 35: Fire Brigade 18
White & Poppe 12, 13
Windovers 33